Three-Way Ratios

Example

Find the ratio of squares to circles.

- Find the number of each item. 4 squares and 3 circles
- You can express the ratio three ways:

as a fraction	using a colon	using the word "to"
$\frac{4}{3}$	4:3	4 to 3

So, the ratio of squares to circles is $\frac{4}{3}$, **4:3**, or **4 to 3**.

Use the picture below. Find the ratio of:

1. houses to buses.
2. buses to dogs.
3. cars to houses.
4. dogs to houses.
5. cats to cars.
6. vehicles to houses.
7. cars to buses.
8. dogs to buses.
9. pets to buses.
10. pets to vehicles.
11. pets to houses.
12. cars to dogs.

Answer Box

A	B	C	D	E	F
3 to 5	1 to 4	$\frac{4}{2}$	2:3	$\frac{4}{3}$	$\frac{3}{7}$
G	**H**	**I**	**J**	**K**	**L**
4 to 5	3 to 3	7:5	3:2	$\frac{5}{3}$	2 to 5

Objective: Name a ratio to compare two quantities from a picture or drawing.

Riding Ratios

Example

Find the ratio of windows to doors.

- The first item is the numerator. number of windows ⟶ $\frac{(4 + 4)}{(1 + 1)}$
- The second item is the denominator. number of doors ⟶

So, the ratio of windows to doors is $\frac{8}{2}$, or **8:2**, or **8 to 2**.

Use the items below. Find the ratio of:

1. bicycles to bicycle wheels.
2. skateboards to skateboard wheels.
3. tricycles to tricycle wheels.
4. skateboards to all vehicles.
5. all wheels to all cycles.
6. bicycles to tricycles.
7. skateboards to bicycles.
8. tricycles to skateboards.
9. all cycles to skateboards.
10. tricycles to all wheels.
11. all wheels to skateboards.
12. skateboards to tricycles.

Answer Box

A	B	C	D	E	F
26 to 7	$\frac{26}{2}$	2:3	4:26	$\frac{2}{4}$	4 to 2

G	H	I	J	K	L
7:2	$\frac{4}{12}$	$\frac{3}{6}$	2:9	2 to 8	$\frac{3}{4}$

 Objective: Name a ratio to compare two quantities from a picture or drawing.

Tasty Ratios

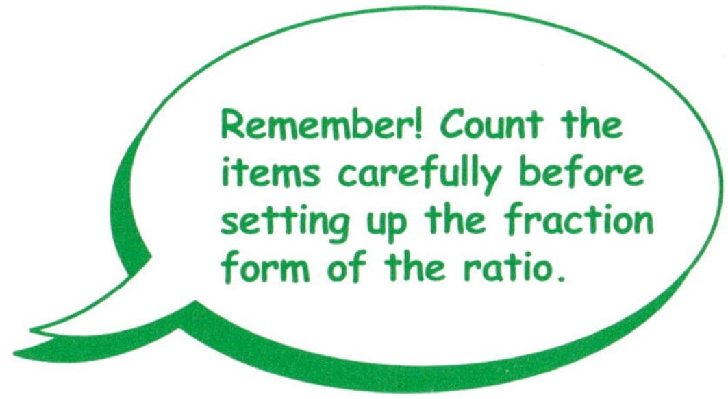

Find the ratio of:

1. rectangular pizza slices to round pizza slices.
2. round pizza slices to drinks.
3. drinks to sandwiches.
4. sandwich slices to pizza slices.
5. cups to cans.
6. round pizza slices to sandwich slices.
7. round pizza slices to rectangular pizza slices.
8. sandwiches to pizzas.
9. sandwich slices to pizzas.
10. drinks to sandwich slices.
11. pizza slices to sandwich slices.
12. cans to drinks.

Answer Box

A	B	C	D	E	F
$\frac{12}{8}$	$\frac{8}{12}$	$\frac{8}{10}$	$\frac{8}{5}$	$\frac{20}{10}$	$\frac{5}{2}$
G	**H**	**I**	**J**	**K**	**L**
$\frac{5}{10}$	$\frac{10}{20}$	$\frac{2}{5}$	$\frac{2}{2}$	$\frac{10}{2}$	$\frac{3}{2}$

Objective: Name a ratio as a fraction to compare two quantities from a picture or drawing.

Sporty Ratios

Example

Find the ratio of gloves to balls.

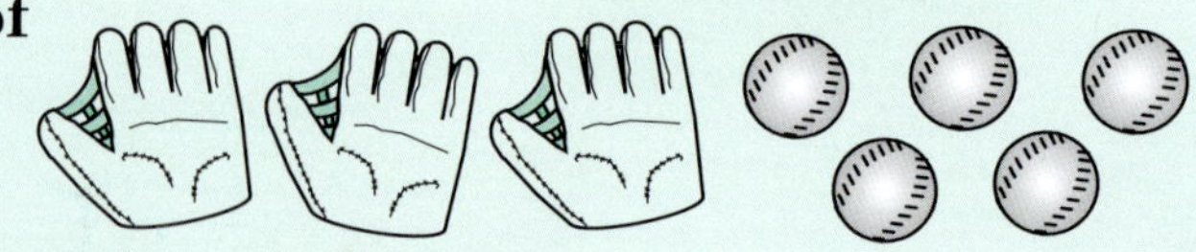

gloves to balls = gloves:balls = 3:5

So, the ratio of gloves to balls is **3:5**.

Use the sports equipment below. Find the ratio of:

1. hockey equipment to baseball equipment.
2. football equipment to hockey equipment.
3. baseball equipment to football equipment.
4. baseball equipment to hockey equipment.
5. balls to sticks.
6. pucks to helmets.
7. gloves to sticks.
8. pucks to balls.
9. bats to balls.
10. bats to helmets.
11. helmets to balls.
12. footballs to baseballs.

Answer Box

A	B	C	D	E	F
7:6	1:3	1:2	1:6	1:4	4:2
G	**H**	**I**	**J**	**K**	**L**
4:4	4:6	6:5	5:6	6:4	5:7

Objective: Name a ratio using a colon to compare two quantities from a picture or drawing.

Keep It Simple

Find the ratio in simplest form.

1. math texts to social studies texts
2. English texts to math texts
3. math texts to all textbooks
4. all textbooks to social studies texts
5. English texts to social studies texts
6. social studies texts to math texts
7. novels to all texts
8. biographies to novels
9. biographies to English texts
10. non-texts to all texts
11. novels to math texts
12. social studies texts to novels

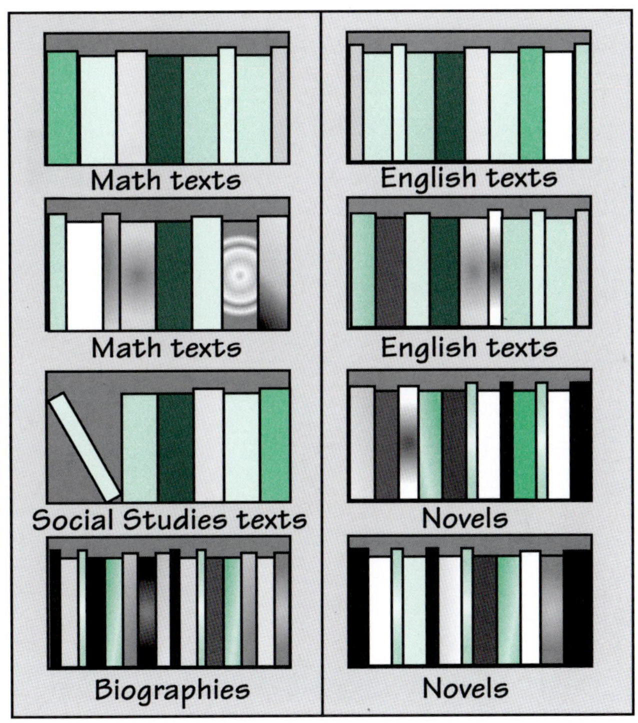

Answer Box

A	B	C	D	E	F
$\frac{2}{3}$	$\frac{3}{8}$	$\frac{8}{21}$	$\frac{4}{5}$	$\frac{20}{21}$	$\frac{7}{1}$
G	**H**	**I**	**J**	**K**	**L**
$\frac{5}{4}$	$\frac{1}{4}$	$\frac{10}{3}$	$\frac{3}{2}$	$\frac{8}{3}$	$\frac{4}{7}$

Objective: Name a ratio in simplest form by dividing both terms by the same nonzero number.

Equality I

Example

Find an equal ratio.

$\frac{5}{6}$

Multiply both terms of the ratio by the same number.
Any number not equal to 0 can be used.

$$\frac{5}{6} \times \frac{4}{4} = \frac{20}{24}$$

So, a ratio equal to $\frac{5}{6}$ is $\mathbf{\frac{20}{24}}$.

Use the shapes below. Find an equal ratio.

1. rectangles to triangles
2. polygons to ovals
3. circles to polygons
4. ovals to polygons
5. circles and ovals to polygons
6. triangles to circles
7. rectangles to all shapes
8. polygons to circles and rectangles
9. circles to triangles
10. triangles to rectangles and ovals
11. ovals to circles
12. all shapes to polygons

Answer Box

A	B	C	D	E	F
$\frac{6}{4}$	$\frac{12}{24}$	$\frac{12}{6}$	$\frac{8}{28}$	$\frac{18}{15}$	$\frac{8}{12}$
G	**H**	**I**	**J**	**K**	**L**
$\frac{24}{18}$	$\frac{15}{9}$	$\frac{8}{36}$	$\frac{10}{12}$	$\frac{28}{12}$	$\frac{12}{14}$

Objective: Name an equal ratio for a given ratio by multiplying each term by the same nonzero number.

Equality II

Find an equal ratio.

1. 24 books to 4 shelves
2. 25 pairs of pants to 5 hangers
3. 27 books to 54 backpacks
4. 8 bulbs in 4 lamps
5. 16 legs on 4 chairs
6. 48 tires on 12 cars
7. 7 baby carriages need 28 wheels
8. 4 tables need 24 chairs
9. 15 plants on 9 shelves
10. 40 pencils to 28 pens
11. 9 tricycles have 27 wheels
12. 20 computers use 20 mice

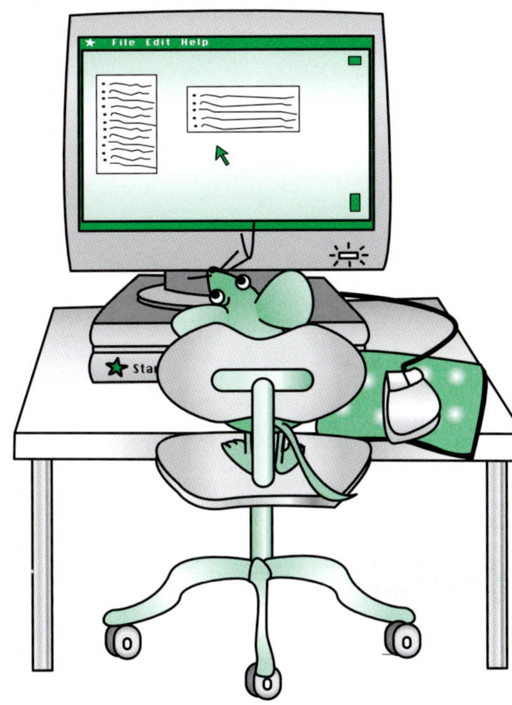

Answer Box

A	B	C	D	E	F
$\frac{12}{2}$	4 to 2	5:3	$\frac{1}{2}$	5 to 1	$\frac{12}{3}$
G	**H**	**I**	**J**	**K**	**L**
1:4	10:7	8 to 2	1:3	$\frac{2}{2}$	1 to 6

Objective: Name an equal ratio for a given ratio by dividing each term by the same nonzero number.

What's the Rate?

	Bill	Susan	Fred	Kathy	Rodrigo
Books Read	4	6	5	3	5
Reading Time (in hours)	35	50	40	27	35
Reports Written	3	2	3	1	3
Writing Time (in hours)	12	9	15	4	18

CONSIDER THIS

Reading rate = $\frac{\text{books read}}{\text{reading time}}$

Writing rate = $\frac{\text{reports written}}{\text{writing time}}$

Find the:

1. reading rate for Susan.
2. writing rate for Rodrigo.
3. writing rate for Bill.
4. reading rate for Fred.
5. writing rate for Kathy.
6. reading rate for Bill.
7. reading rate for Kathy.
8. writing rate for Susan.

Which student had a:

9. reading rate of $\frac{\text{5 books}}{\text{35 h}}$.
10. writing rate of $\frac{\text{3 reports}}{\text{15 h}}$.
11. writing rate of $\frac{\text{2 reports}}{\text{9 h}}$.
12. reading rate of $\frac{\text{4 books}}{\text{35 h}}$.

Answer Box

A	B	C	D	E	F
$\frac{\text{2 reports}}{\text{9 h}}$	Bill	Susan	$\frac{\text{1 report}}{\text{4 h}}$	Fred	$\frac{\text{3 books}}{\text{27 h}}$
G	**H**	**I**	**J**	**K**	**L**
$\frac{\text{3 reports}}{\text{18 h}}$	$\frac{\text{4 books}}{\text{35 h}}$	$\frac{\text{3 reports}}{\text{12 h}}$	Rodrigo	$\frac{\text{6 books}}{\text{50 h}}$	$\frac{\text{5 books}}{\text{40 h}}$

 Objective: Express a given situation using a rate.

A Unit Is One

Example

Find the unit rate.

28 days in 4 weeks

- Find the ratio as a fraction. $\frac{28}{4}$
- Divide the numerator and denominator by their GCF. $\frac{28}{4} \div \frac{4}{4} = \frac{7}{1}$

So, the unit rate for 28 days in 4 weeks is **7 days/week**.

Find the unit rate.

1. $75.00 earned in 3 h
2. 18 packages wrapped in 2 h
3. $48.00 spent on 3 copies of a book
4. 60 pictures on 5 pages
5. 36 servings in 2 boxes
6. $3.75 paid for 3 tolls
7. 48 packages delivered in 3 h
8. 96 servings in 6 boxes
9. $7.50 paid for 5 tolls
10. 35 lamp posts on 5 streets
11. 18 pictures on 2 pages
12. 96 parking spaces on 12 streets

Answer Box

A	B	C	D	E	F
$16.00/ book	$1.50/toll	$25.00/h	12/page	$1.25/toll	9/h
G	**H**	**I**	**J**	**K**	**L**
18/box	9/page	16/h	7/street	8/street	16/box

Objective: Name a unit rate for a given rate by forming an equal ratio.

Shopping Is Work

Example

Find the unit price.

Divide to find the unit price of each item, rounded to the nearest cent.

$$10\overline{)\$2.19}\quad = \$0.219 \qquad 14\overline{)\$2.39}\quad = \$0.170$$

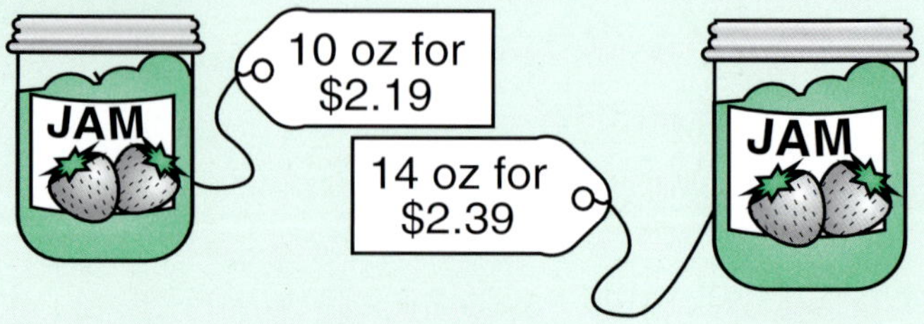

So, $2.19 for 10 oz = **$0.22 per oz** and $2.39 for 14 oz = **$0.17 per oz**.

Find the unit price.

1. apples — 12 for $1.39
2. pears — 5 for $1.45
3. gum — 2 packs for $1.58
4. markers — 9 for $9.18
5. comics — 3 packs for $2.97
6. comics — 6 packs for $6.90
7. ice cream — 10 pops for $14.50
8. snacks — 9 oz for $1.71
9. pens — 8 for $1.09
10. seeds — 3 packs for $4.05
11. pancake syrup — 8 fl oz for $1.99
12. tomato juice — 64 fl oz for $3.00

Answer Box

A	B	C	D	E	F
$1.35	$1.45	$0.05	$0.29	$0.25	$0.99
G	**H**	**I**	**J**	**K**	**L**
$1.02	$0.12	$1.15	$0.19	$0.79	$0.14

Objective: Use unit rates to determine the unit price.

Shopping Takes Time

Example

Choose the better buy.

3 pens for $2.95 or 5 pens for $5.10

Sometimes you can estimate to find the better buy.
If 3 pens cost $2.95, then each pen costs less than $1.00.
If 5 pens cost $5.10, then each pen costs more than $1.00.

So, **3 for $2.95** is the better buy.

Choose the better buy.

1. 3 lb for $1.49 or 5 lb for $2.79
2. 10 oz for $0.99 or 9 oz for $1.71
3. 7 oz for $0.59 or 15 oz for $0.89
4. 2 pieces for $0.81 or 12 pieces for $2.69
5. 9 pieces for $1.71 or 12 pieces for $2.69
6. 14 CDs for $139.30 or 9 CDs for $92.50
7. 3 movies for $59.50 or 2 movies for $42.10
8. 7 oz for $0.59 or 10 oz for $0.99
9. 2 lb for $0.81 or 3 lb for $1.49
10. 15 oz for $0.89 or 32 oz for $1.59
11. 7 movies for $175.99 or 2 movies for $42.10
12. 9 CDs for $92.50 or 3 CDs for $59.50

Answer Box

A	B	C	D	E	F
12/$2.69	3 lb/$1.49	15 oz/$0.89	2 lb/$0.81	14/$139.30	10 oz/$0.99
G	**H**	**I**	**J**	**K**	**L**
3/$59.50	32 oz/$1.59	9/$1.71	2/$42.10	9/$92.50	7 oz/$0.59

Objective: Compare prices using unit rates to determine the better buy.

Problem Solving: Applying Rates

U-Drive Them Car Rental				
Type of Car	**Daily Rate**	**Miles Over 50**	**Weekly Rate**	**Miles Over 400**
Compact	$26.99	$0.15/mi	$167.93	$0.15/mi
Mid-Size	$31.49	$0.15/mi	$192.43	$0.15/mi
Luxury	$39.99	$0.20/mi	$241.43	$0.20/mi
Additional charge if not returned within 24 h: $7.00 per hour for all size cars. Senior Citizen Discount: $10.00 off the total for any rental over $50.00.				

Use the data in the table to answer the question.

1. Matt picks up a mid-size car at 9:00 A.M. Monday and returns it at 7:30 A.M. Tuesday. If he drove a total of 35 mi, what was his cost?

2. The agency rents 7 compact cars for 1 day. What is the total they collect if none of the cars are driven over 50 mi?

3. How much does Kara pay if she rents a luxury car for one week and drives it 750 mi?

4. How much more does Yoshi pay if she rents a mid-size car for 1 day and drives it a total of 99 mi than if she drove the same car for only 44 mi?

5. How much does Hector pay for a mid-size car if he drives it under 50 mi per day and keeps it for 3 full days and 4 h on the day he returns it?

6 How much does a senior citizen pay for a luxury car for a 2-day rental and drives it 200 mi?

7 How much more does Sari pay to rent a compact car for 4 days and drive 300 mi than a senior citizen renting the same car and driving the same mileage?

8 How much less does it cost Harry to rent a compact car by the day for 7 days than a mid-size car by the week for one week, assuming he drives under 50 mi per day or under 400 mi per week?

9 How much more would Tony pay for a 4-week rental of a luxury car than for a 4-week rental of a mid-size car assuming he drives under 400 mi per week?

How much more do you pay if you rent the car by the day for 7 days instead of paying for a full week's rental?

10 Compact car

11 Mid-size car

12 Luxury car

Answer Box

A	B	C	D	E	F
$38.50	$3.50	$21	$7.35	$122.47	$28
G	**H**	**I**	**J**	**K**	**L**
$99.98	$188.93	$31.49	$10.00	$311.43	$196.00

Objective: Solve a problem by using a rate.

Thinking Proportionally

Example

Choose the fraction that completes the proportion.

$\frac{8}{12}$ $\frac{10}{14}$ or $\frac{6}{8}$ or $\frac{16}{24}$

Test each fraction.

$\frac{8}{12} \neq \frac{10}{14}$ There is no common factor by which you can multiply $\frac{8}{12}$ to get $\frac{10}{14}$.

$\frac{8}{12} \neq \frac{6}{8}$ There is no common factor by which you can divide $\frac{8}{12}$ to get $\frac{6}{8}$.

$\frac{8}{12} = \frac{16}{24}$ When 8 and 12 are each multiplied by 2, the result is $\frac{16}{24}$.

So, $\frac{8}{12}$ and $\mathbf{\frac{16}{24}}$ form a proportion.

Choose the fraction that completes the proportion.

1. $\frac{7}{8}$ $\frac{15}{24}$ or $\frac{10}{12}$ or $\frac{14}{16}$
2. $\frac{5}{7}$ $\frac{14}{16}$ or $\frac{15}{21}$ or $\frac{15}{24}$
3. $\frac{3}{5}$ $\frac{6}{15}$ or $\frac{14}{20}$ or $\frac{12}{20}$
4. $\frac{4}{9}$ $\frac{12}{27}$ or $\frac{12}{20}$ or $\frac{8}{24}$
5. $\frac{1}{3}$ $\frac{8}{20}$ or $\frac{8}{24}$ or $\frac{7}{8}$
6. $\frac{7}{10}$ $\frac{14}{20}$ or $\frac{12}{20}$ or $\frac{21}{28}$
7. $\frac{20}{28}$ $\frac{5}{7}$ or $\frac{6}{7}$ or $\frac{7}{8}$
8. $\frac{30}{40}$ $\frac{4}{5}$ or $\frac{3}{4}$ or $\frac{5}{6}$
9. $\frac{21}{24}$ $\frac{7}{9}$ or $\frac{7}{8}$ or $\frac{3}{4}$
10. $\frac{36}{45}$ $\frac{4}{5}$ or $\frac{5}{7}$ or $\frac{3}{4}$
11. $\frac{10}{16}$ $\frac{3}{4}$ or $\frac{15}{24}$ or $\frac{4}{5}$
12. $\frac{10}{12}$ $\frac{8}{24}$ or $\frac{7}{8}$ or $\frac{5}{6}$

Answer Box

A	B	C	D	E	F
$\frac{15}{24}$	$\frac{12}{20}$	$\frac{4}{5}$	$\frac{12}{27}$	$\frac{3}{4}$	$\frac{14}{20}$
G	**H**	**I**	**J**	**K**	**L**
$\frac{7}{8}$	$\frac{8}{24}$	$\frac{14}{16}$	$\frac{5}{7}$	$\frac{15}{21}$	$\frac{5}{6}$

 Objective: Complete a proportion from a set of given ratios.

Product Crossing

Find the cross product.

1. $\frac{5}{8} = \frac{15}{24}$
2. $\frac{9}{10} = \frac{27}{30}$
3. $\frac{7}{11} = \frac{21}{33}$
4. $\frac{6}{14} = \frac{3}{7}$
5. $\frac{14}{35} = \frac{2}{5}$
6. $\frac{3}{8} = \frac{12}{32}$
7. $\frac{3}{7} = \frac{9}{21}$
8. $\frac{4}{9} = \frac{16}{36}$
9. $\frac{1}{8} = \frac{3}{24}$
10. $\frac{5}{12} = \frac{15}{36}$
11. $\frac{3}{5} = \frac{12}{20}$
12. $\frac{7}{10} = \frac{21}{30}$

Answer Box

A	B	C	D	E	F
231	24	42	120	270	96
G	**H**	**I**	**J**	**K**	**L**
60	70	63	180	210	144

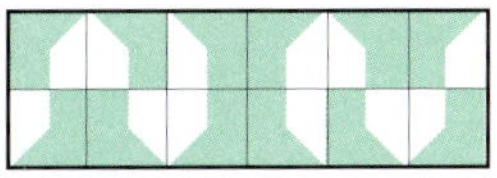

Find the cross product.

1. $\frac{9}{10} = \frac{63}{70}$
2. $\frac{3}{11} = \frac{12}{44}$
3. $\frac{15}{18} = \frac{5}{6}$
4. $\frac{4}{13} = \frac{8}{26}$
5. $\frac{7}{15} = \frac{21}{45}$
6. $\frac{1}{3} = \frac{21}{63}$
7. $\frac{15}{25} = \frac{3}{5}$
8. $\frac{6}{7} = \frac{18}{21}$
9. $\frac{27}{36} = \frac{3}{4}$
10. $\frac{2}{9} = \frac{6}{27}$
11. $\frac{5}{12} = \frac{15}{36}$
12. $\frac{12}{15} = \frac{4}{5}$

Answer Box

A	B	C	D	E	F
75	54	126	60	180	315
G	**H**	**I**	**J**	**K**	**L**
270	104	132	63	90	108

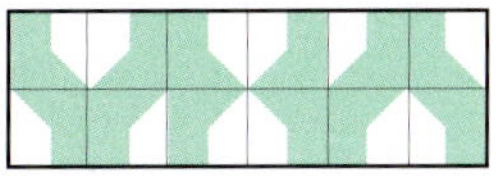

Objective: Identify the cross product given a proportion.

What's Missing?

Find the missing number.

1. $\frac{3}{4} = \frac{\blacksquare}{12}$
2. $\frac{12}{16} = \frac{\blacksquare}{4}$
3. $\frac{18}{24} = \frac{9}{\blacksquare}$
4. $\frac{14}{21} = \frac{\blacksquare}{9}$
5. $\frac{9}{15} = \frac{\blacksquare}{35}$
6. $\frac{8}{10} = \frac{\blacksquare}{25}$
7. $\frac{8}{15} = \frac{\blacksquare}{45}$
8. $\frac{3}{12} = \frac{9}{\blacksquare}$
9. $\frac{\blacksquare}{25} = \frac{4}{10}$
10. $\frac{12}{18} = \frac{18}{\blacksquare}$
11. $\frac{7}{9} = \frac{\blacksquare}{63}$
12. $\frac{10}{\blacksquare} = \frac{15}{21}$

Answer Box

A	B	C	D	E	F
24	9	10	12	14	3
G	**H**	**I**	**J**	**K**	**L**
6	27	21	49	20	36

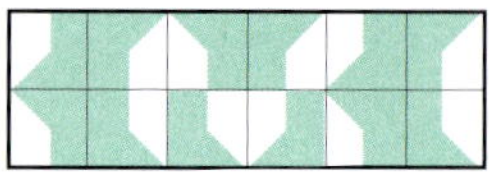

Find the missing number.

1. $\frac{3}{\blacksquare} = \frac{9}{24}$
2. $\frac{15}{18} = \frac{25}{\blacksquare}$
3. $\frac{14}{26} = \frac{21}{\blacksquare}$
4. $\frac{\blacksquare}{16} = \frac{5}{8}$
5. $\frac{21}{24} = \frac{\blacksquare}{16}$
6. $\frac{16}{\blacksquare} = \frac{12}{15}$
7. $\frac{20}{\blacksquare} = \frac{12}{30}$
8. $\frac{\blacksquare}{30} = \frac{9}{15}$
9. $\frac{\blacksquare}{20} = \frac{18}{30}$
10. $\frac{40}{\blacksquare} = \frac{16}{18}$
11. $\frac{9}{14} = \frac{27}{\blacksquare}$
12. $\frac{8}{18} = \frac{16}{\blacksquare}$

Answer Box

A	B	C	D	E	F
18	36	8	10	14	42
G	**H**	**I**	**J**	**K**	**L**
20	30	50	45	12	39

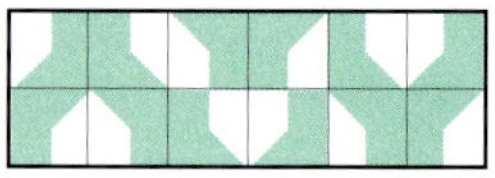

Objective: Find the missing term in a proportion.

Similarities

Use the similar figures. Find the length of the missing side.

1. *m*
2. *n*
3. *p*
4. *q*
5. *r*
6. *s*
7. *t*
8. *u*
9. *v*
10. *w*
11. *x*
12. *y*

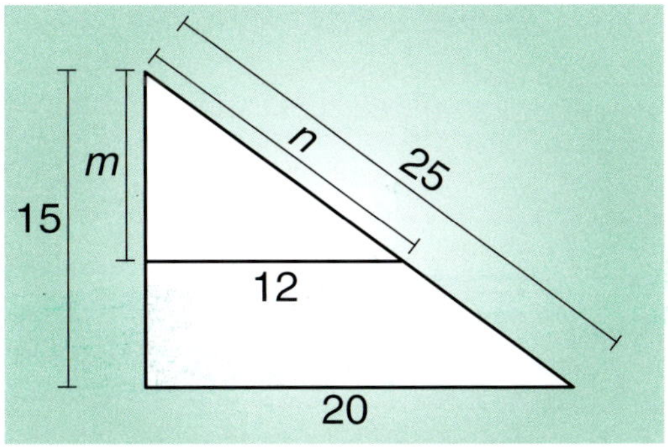

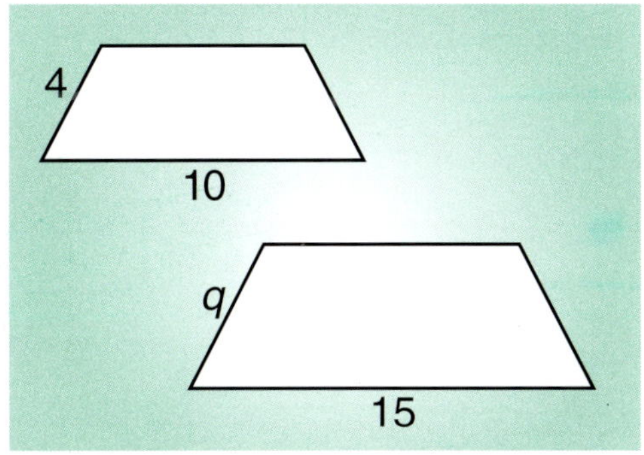

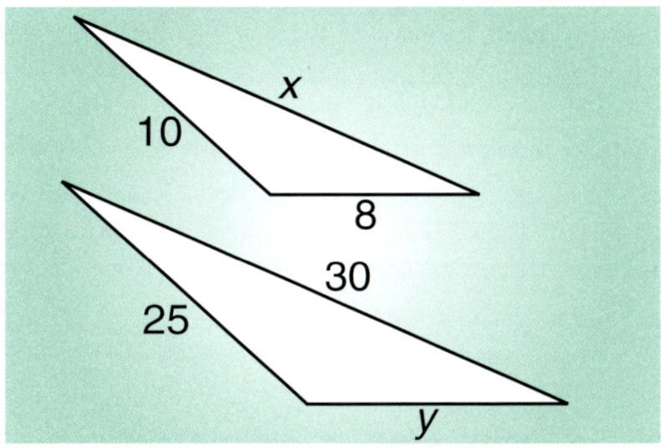

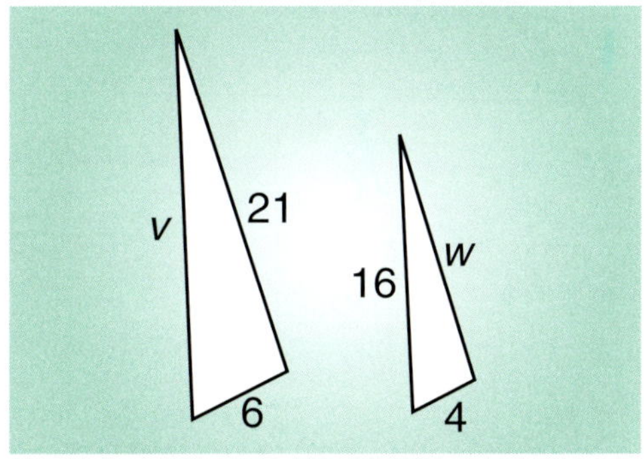

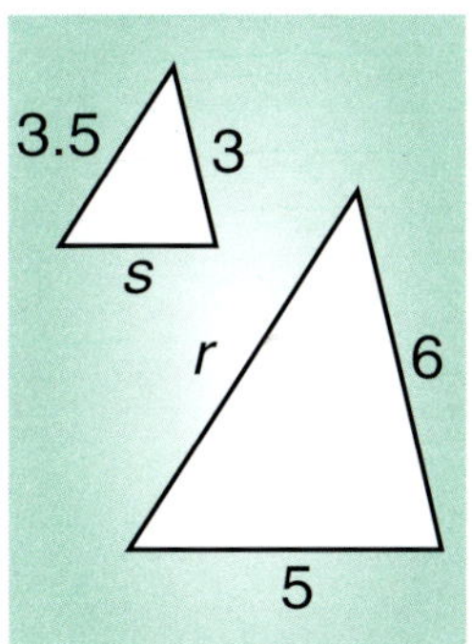

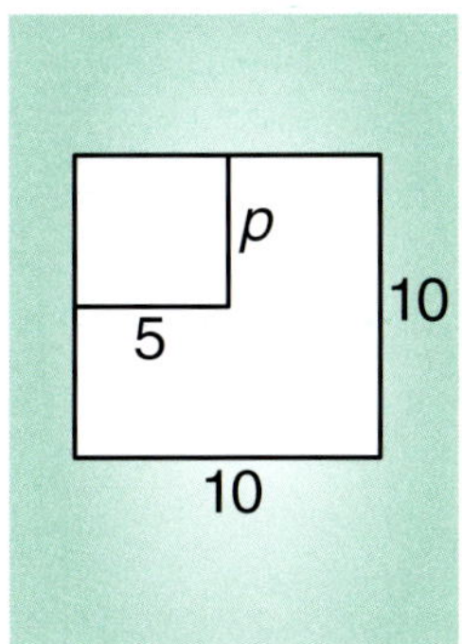

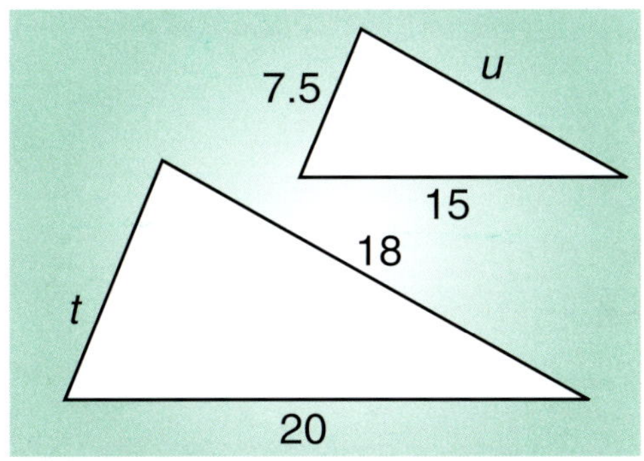

Answer Box

A	B	C	D	E	F
13.5	20	12	7	14	10
G	**H**	**I**	**J**	**K**	**L**
15	2.5	24	5	9	6

Objective: Find the missing measure in similar figures using a proportion.

Look Similar?

Find the length of the missing side.

1. *m*
2. *n*
3. *p*
4. *q*
5. *r*
6. *s*
7. *t*
8. *u*
9. *v*
10. *w*
11. *x*
12. *y*

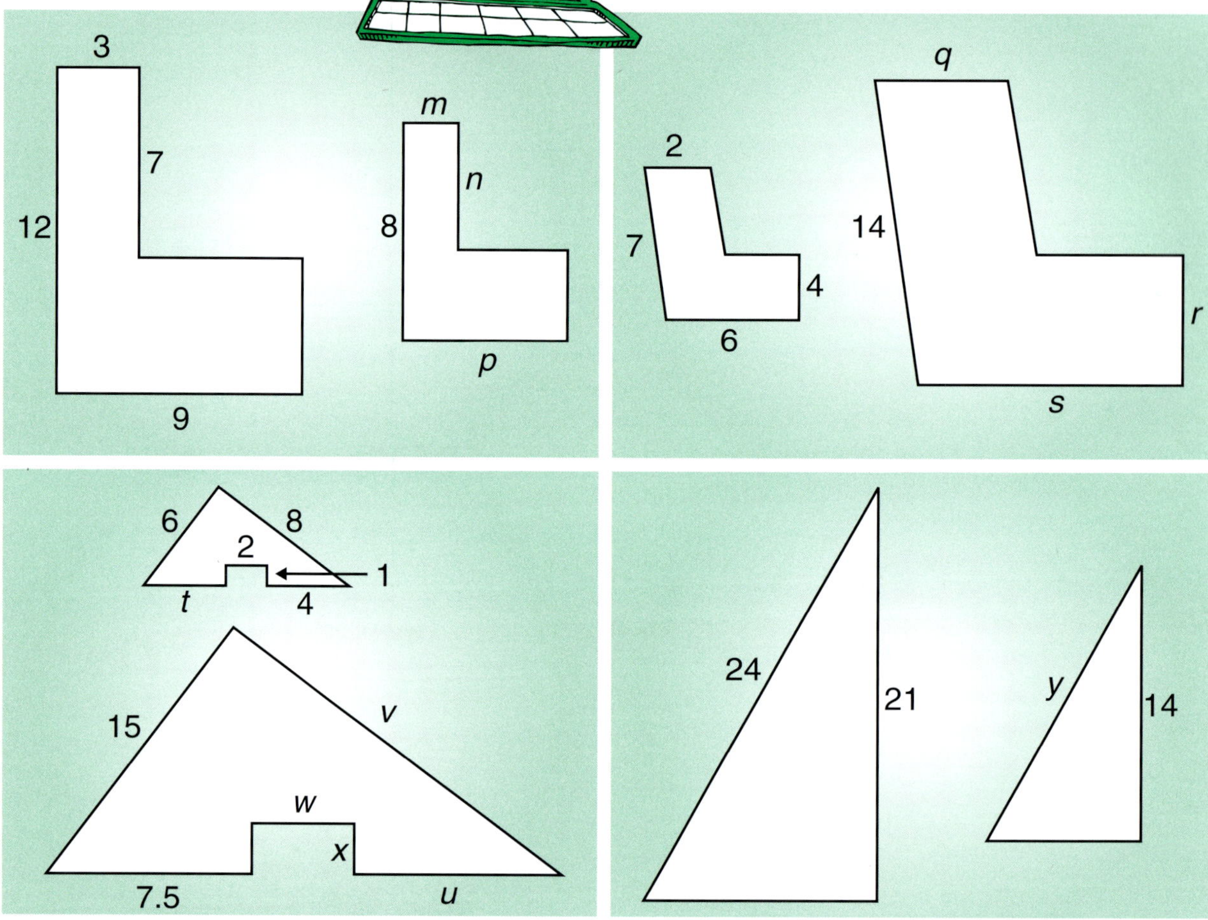

Answer Box

A	B	C	D	E	F
6	2.5	4	5	12	10
G	**H**	**I**	**J**	**K**	**L**
8	20	3	2	16	4.7

 Objective: Find the missing measure in similar figures using a proportion.

Drawn to Scale

Example

Find the scale length.

actual height of tree: 18 ft
scale of drawing: 1 in.:3 ft

Set up and solve a proportion using the fact that the actual tree and its scale drawing are similar figures.

$$\frac{1 \text{ in.}}{3 \text{ ft}} = \frac{\blacksquare \text{ in.}}{18 \text{ ft}}$$

$$3 \times \blacksquare = 18$$

$$\blacksquare = \frac{18}{3} = 6$$

So, the scale length (the height of the drawing) is **6 in.**

Find the scale length.

1	*m*	2	*n*
3	*p*	4	*q*
5	*r*	6	*s*
7	*t*	8	*u*
9	*v*	10	*w*
11	*x*	12	*y*

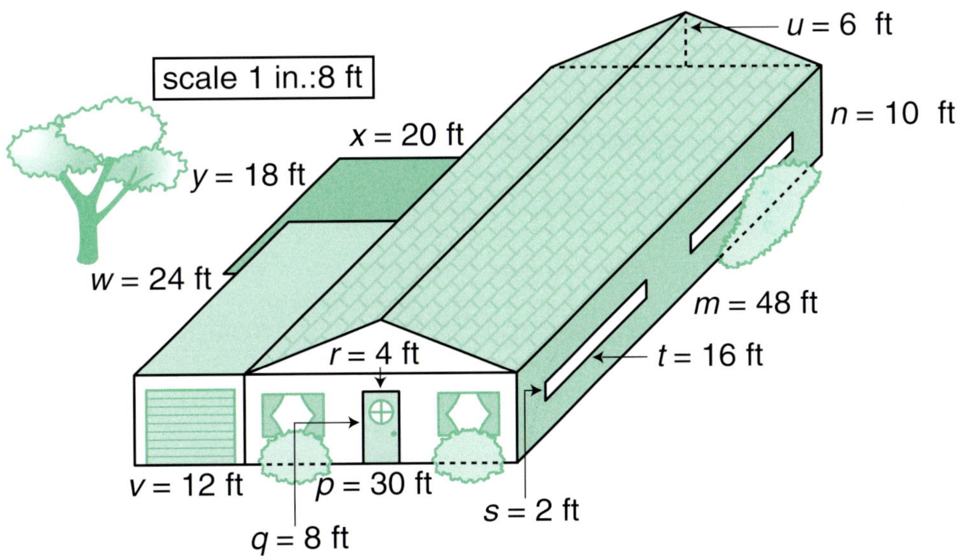

Answer Box

A	B	C	D	E	F
6 in.	1 in.	2.25 in.	0.75 in.	3.75 in.	1.5 in.
G	**H**	**I**	**J**	**K**	**L**
2 in.	3 in.	0.25 in.	1.25 in.	2.5 in.	0.5 in.

Objective: Find the scale length given the actual length and scale ratio.

Jump on the Scale!

Example

Find the actual length.

tree scale height: 7 in. scale: 1 in.:3 ft

- Write and solve a proportion.

$$\frac{7 \text{ in.}}{\blacksquare \text{ ft}} = \frac{1 \text{ in.}}{3 \text{ ft}}$$

$$\blacksquare = 21$$

So, the actual height of the tree is **21 ft**.

Find the actual length.

1. key scale length: 2 cm scale: 1 cm:5 cm
2. flagpole scale length: 5 in. scale: 1 in.:4 ft
3. trash can scale length: 6 cm scale: 1 cm:8 cm
4. book scale length: 2.5 in. scale: 1 in.:4 in.
5. desk scale length: 5.5 cm scale: 1 cm:1.5 ft
6. chair scale length: 4 in. scale: 1 in.:0.75 ft
7. window scale length: 6 cm scale: 1 cm:0.5 m
8. airplane scale length: 6.5 in. scale: 1 in.:7 ft
9. building scale length: 3.8 in. scale: 1 in.:12 ft
10. room scale length: 3.5 in. scale: 1 in.:6 ft
11. tree scale length: 7.5 cm scale: 1 cm:2.5 m
12. wall scale length: 6.2 cm scale: 1 cm:1.5 m

Answer Box

A	B	C	D	E	F
10 in.	10 cm	45.6 ft	8.25 ft	3 ft	20 ft
G	**H**	**I**	**J**	**K**	**L**
21 ft	3 m	18.75 m	48 cm	45.5 ft	9.3 m

 Objective: Find the actual length given the scale length and scale ratio.

Actual or Scale?

Find the scale length.

1	horse	actual height: 5.5 ft	scale: 1 in.:2.5 ft
2	student	actual height: 6 ft	scale: 1 in.:1.5 ft
3	refrigerator	actual height: 5 ft	scale: 1 in.:2 ft
4	tent	actual height: 8 ft	scale: 1 in.:2.5 ft
5	TV	actual height: 2.5 ft	scale: 1 in.:0.5 ft
6	bed	actual height: 6.5 ft	scale: 1 in.:2 ft

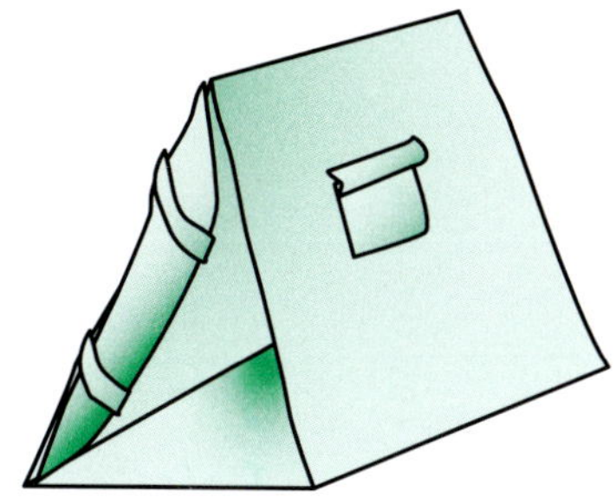

Find the actual length.

7	living room	scale length: 14 cm	scale: 1 cm:0.5 m
8	kitchen	scale length: 10 cm	scale: 1 cm:0.4 m
9	bathroom	scale length: 18 cm	scale: 1 cm:0.2 m
10	hallway	scale length: 12 cm	scale: 1 cm:0.6 m
11	den	scale length: 8.5 cm	scale: 1 cm:0.6 m
12	garage	scale length: 4 cm	scale: 1 cm:1.7 m

Answer Box

A	B	C	D	E	F
6.8 m	4 in.	7.2 m	7 m	5 in.	3.6 m
G	**H**	**I**	**J**	**K**	**L**
3.25 in.	4 m	2.2 in.	3.2 in.	2.5 in.	5.1 m

Objective: Find the scale length given the actual length and scale ratio or find the actual length given the scale length and the scale ratio.

Problem Solving: Using Proportions

Find the answer.

1. A bag of popcorn has $6\frac{1}{2}$ servings. If one serving has 10 g of fat, how many grams of fat are in the entire bag?

2. In one serving of the popcorn, 90 out of the 160 calories come from fat. How many calories are there in 3 servings?

3. How many calories come from fat in 3 servings of popcorn?

4. Cari's car travels 115 mi on 5 gal of gasoline. How many gallons would be needed to drive 46 mi?

5. How many miles can Cari drive on 9 gal?

6. How many gallons of gas would Cari need to drive 575 mi?

7. A recipe that makes 12 muffins uses 2 c of flour, 5 egg whites, 6 egg yolks, and $1\frac{1}{2}$ c of milk. How many muffins can Mark make if he has one dozen eggs?

8. How many cups of milk would Mark need to make three dozen muffins?

9. How many muffins can Mark make with 5 c of flour?

10. Monica earns $16.35 for 3 h of work around the house. How much would Monica earn for 8 h of work?

11. How many hours did Monica work if she earned $90.95?

12. Martha tells her that she earns $12.50 for 2 h of work. At that rate, how much does Martha earn for 7 h of work?

Answer Box

A	B	C	D	E	F
480	$43.75	207	17	65	24
G	**H**	**I**	**J**	**K**	**L**
$4\frac{1}{2}$	25	270	30	$43.60	2

Objective: Solve a problem by using a proportion.

From 0 to 100 in No Time

Find the percent that describes the shaded part.

1 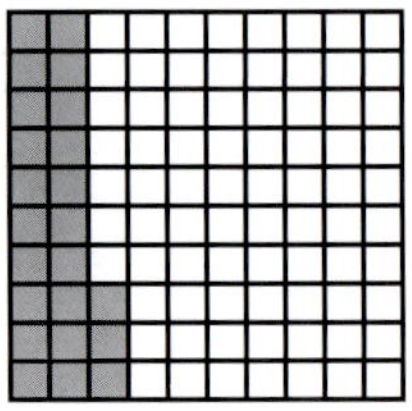

2 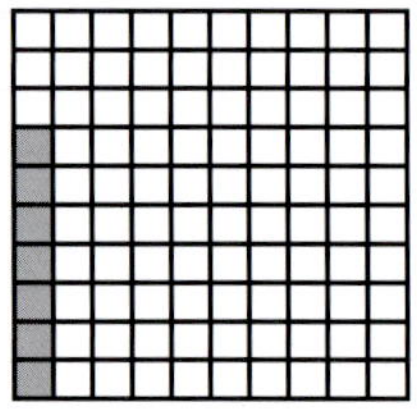

3 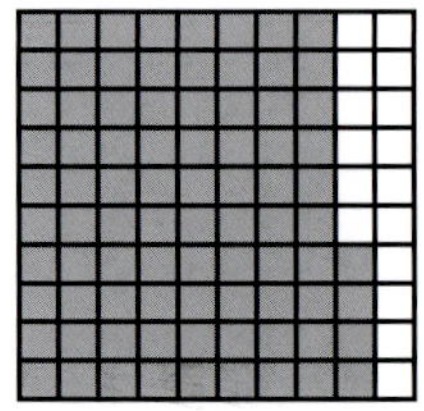

4 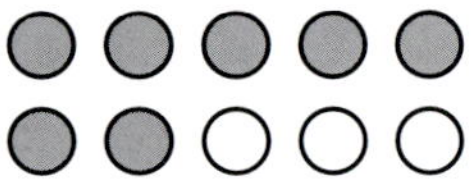

5 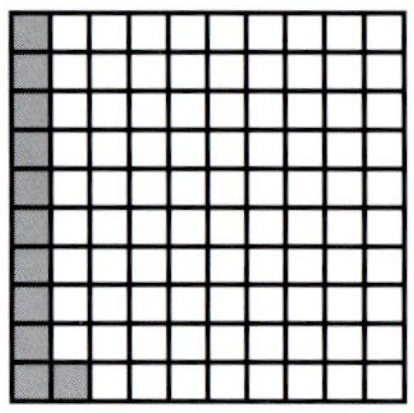

6 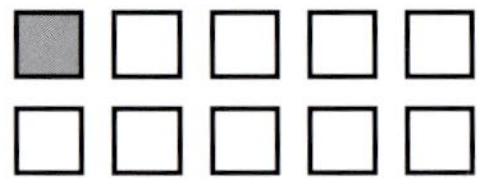

7 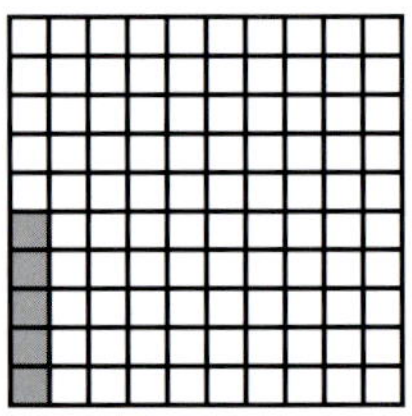

8 

9 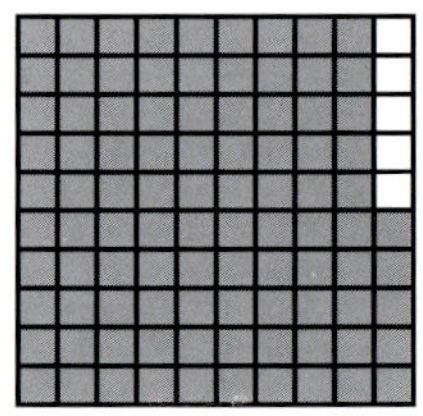

10 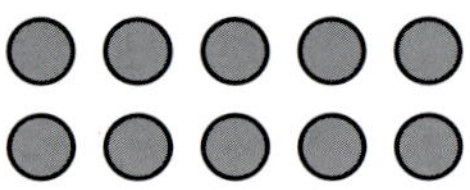

11 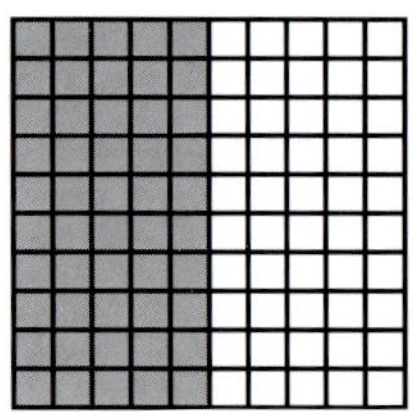

12 

Answer Box

A	B	C	D	E	F
50%	84%	100%	5%	90%	10%
G	**H**	**I**	**J**	**K**	**L**
95%	11%	70%	23%	7%	0%

 Objective: Name a percent from a shaded 10 by 10 grid or from a set of 10 objects.

A Matter of Shading

Find what percent is shaded or unshaded.

1. 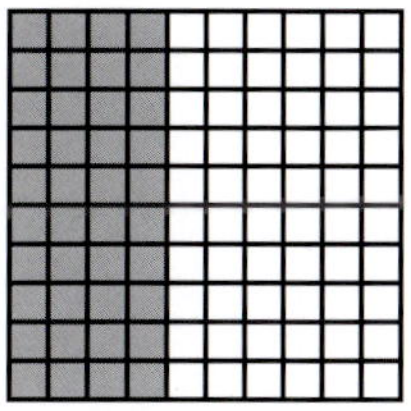
shaded: ■

2. 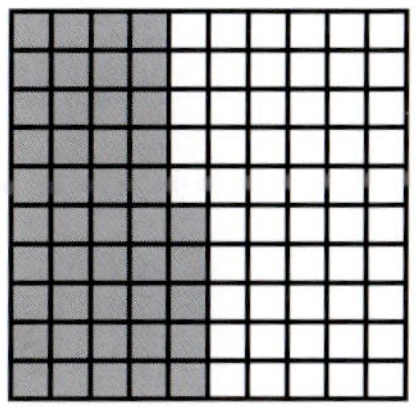
unshaded: ■

3. 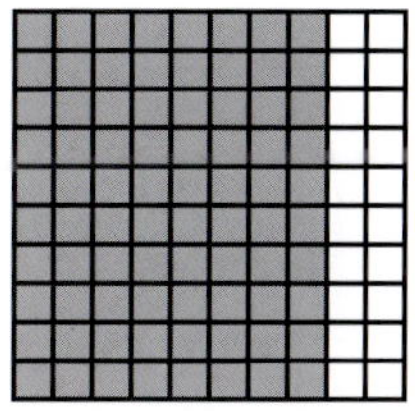
unshaded: ■

4. 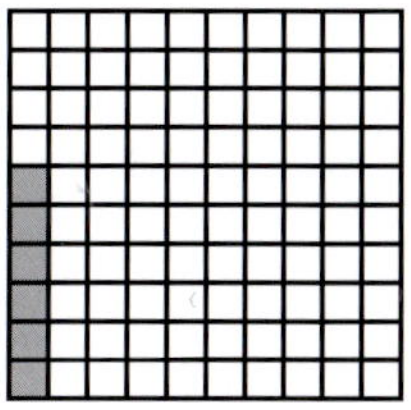
unshaded: ■

5. 
unshaded: ■

6. 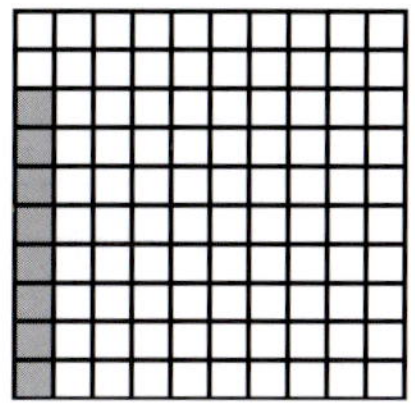
shaded: ■

7. 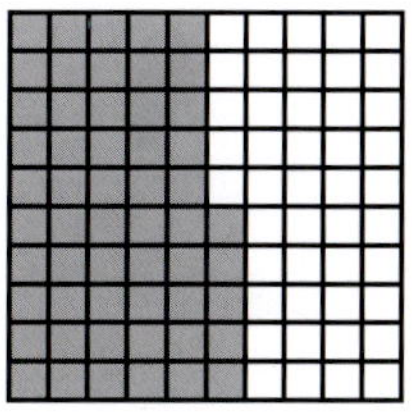
unshaded: ■

8. 
shaded: ■

9. 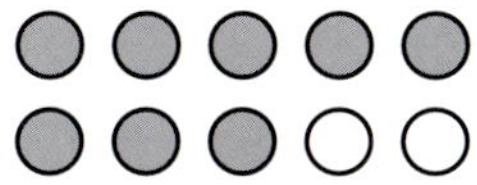
shaded: ■

10. 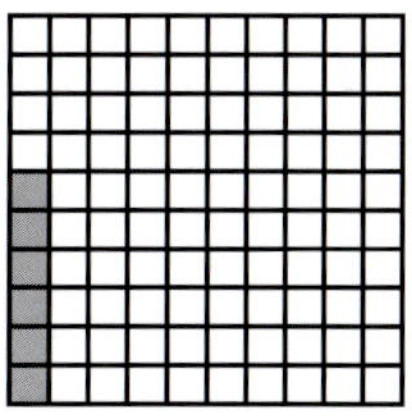
shaded: ■

11. 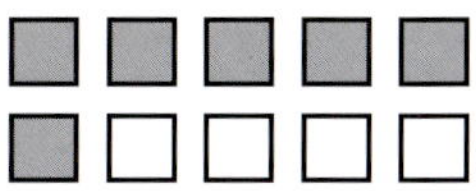
shaded: ■

12. 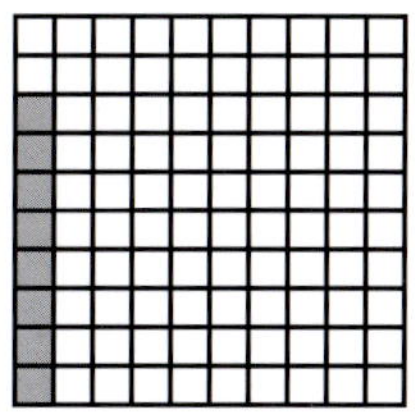
unshaded: ■

Answer Box

A	B	C	D	E	F
45%	6%	92%	10%	90%	60%
G	**H**	**I**	**J**	**K**	**L**
94%	40%	8%	55%	20%	80%

Objective: Name a percent from a shaded 10 by 10 grid or from a set of 10 objects.

Shaded Fractions

Find the fraction that describes the shaded part.

1 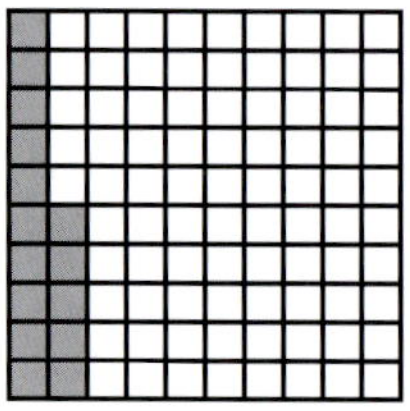

2 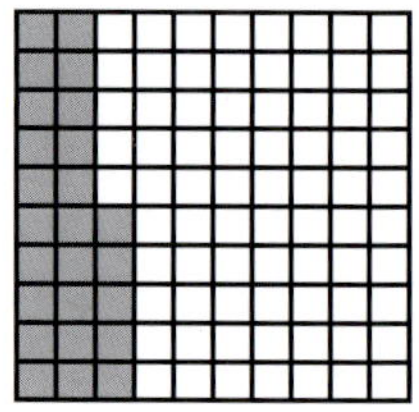

3 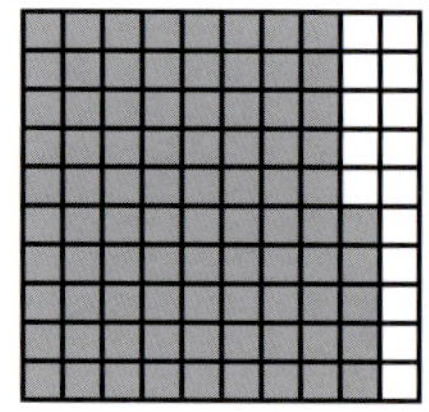

4 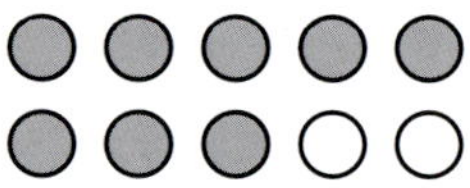

5 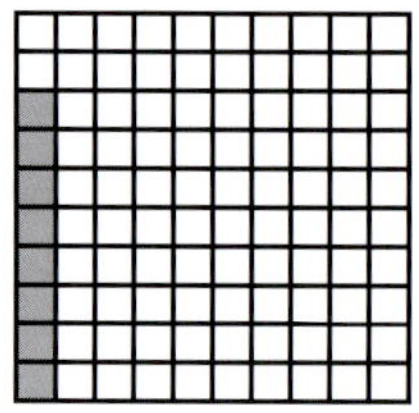

6 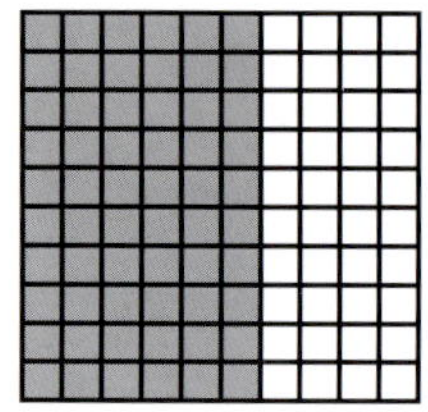

7 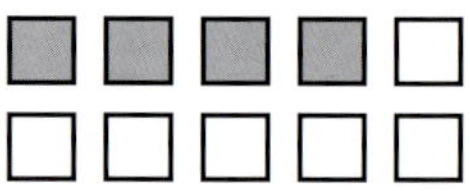

8 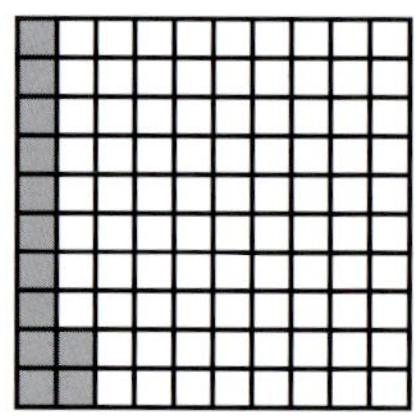

9 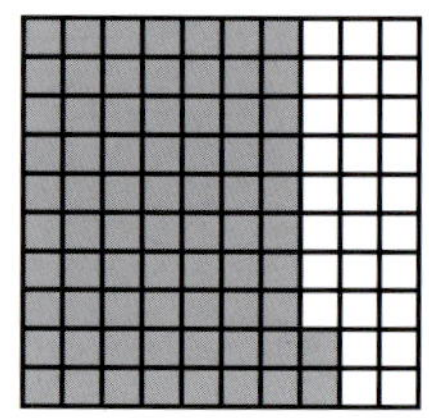

10 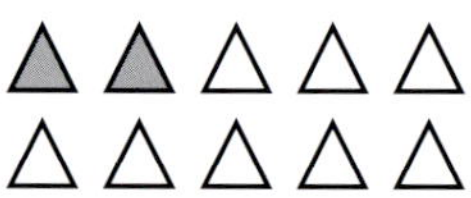

11 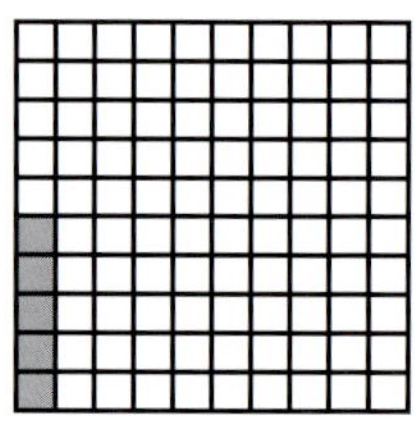

12 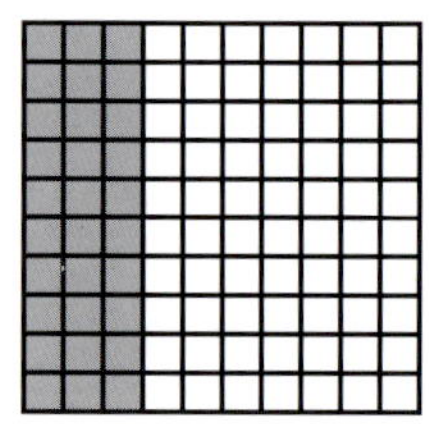

Answer Box

A	B	C	D	E	F
$\frac{2}{5}$	$\frac{1}{5}$	$\frac{1}{20}$	$\frac{2}{25}$	$\frac{3}{10}$	$\frac{3}{25}$
G	**H**	**I**	**J**	**K**	**L**
$\frac{4}{5}$	$\frac{3}{20}$	$\frac{18}{25}$	$\frac{17}{20}$	$\frac{3}{5}$	$\frac{1}{4}$

Objective: Name a percent as a fraction from a shaded 10 x 10 grid or from a set of 10 objects.

Percent of a Number

Example

Find the percent of the number.

25% of 84

- Find the percent as a fraction and as a decimal.

  $25\% = \frac{25}{100} = 0.25$

- Decide whether to use the fraction or decimal form.
- Multiply by the whole number. $0.25 \times 84 = 21$

  So, 25% of 84 = **21**.

Find the percent of the number.

1. 20% of 60
2. 55% of 110
3. 40% of 88
4. 8% of 200
5. 15% of 130
6. 95% of 40
7. 22% of 50
8. 85% of 70
9. 5% of 130
10. 18% of 150
11. The Bulls lost 25% of their 36 games. How many games did the Bulls lose?
12. Carol scored 35% of the Bulls' 40 points. How many points did she scored?

Answer Box

A	B	C	D	E	F
38	60.5	16	12	35.2	6.5
G	**H**	**I**	**J**	**K**	**L**
59.5	14	27	11	9	19.5

Objective: Find the percent of a number using a decimal or fraction for the percent.

More Than Everything?

Find the percent of the number.

1. 90% of 50
2. 150% of 20
3. 125% of 12
4. 80% of 40
5. 200% of 18
6. 120% of 40
7. 60% of 90
8. 130% of 20
9. 250% of 16
10. 220% of 15
11. 125% of 40
12. 160% of 10

Answer Box

A	B	C	D	E	F
40	15	48	26	45	32
G	**H**	**I**	**J**	**K**	**L**
36	50	16	30	54	33

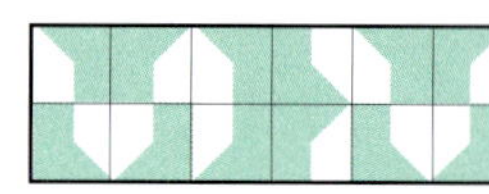

Find the percent of the number.

1. 80% of 70
2. 95% of 120
3. 55% of 86
4. 115% of 60
5. 48% of 80
6. 24% of 90
7. 145% of 20
8. 66% of 55
9. 175% of 36
10. 210% of 40
11. 38% of 125
12. 105% of 84

Answer Box

A	B	C	D	E	F
38.4	47.5	56	69	36.3	88.2
G	**H**	**I**	**J**	**K**	**L**
63	47.3	84	29	21.6	114

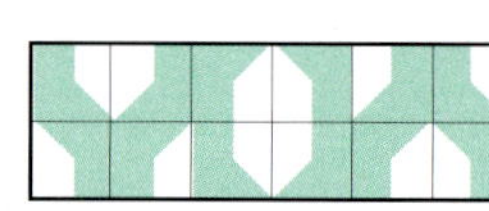

 Objective: Find the percent of a number using a decimal or a fraction for the percent.

It Helps to Be Compatible

Remember! You can change a percent to a fraction by writing the percent over 100 and simplifying.

Example

Find the best estimate.

18% of 82

- Find a fraction that is close to the given percent. — 18% is close to 20%, or $\frac{1}{5}$
- Find a number that is close to and compatible with the given number. — 82 is close to 80.
- Multiply to find an estimate. — $\frac{1}{5} \times 80 = 16$

So, 18% of 82 is about **16**.

Find the best estimate.

1. 49% of 53
2. 78% of 39
3. 91% of 61
4. 53% of 99
5. 9% of 71
6. 26% of 47
7. 68% of 21
8. 32% of 59
9. 48% of 91
10. 21% of 24
11. Tom tells Dave that he is about 70% of the way to his grandmother's house. If the entire trip is 89 mi, about how many miles has he traveled?
12. The gas gauge shows that the tank is about 25% full. It holds 23 gal in all. About how many gallons of gas are in the tank?

Answer Box

A	B	C	D	E	F
25	14	30	6	45	7
G	**H**	**I**	**J**	**K**	**L**
5	50	20	12	63	54

Objective: Find the estimated percent of a number using compatible numbers.

Problem Solving: Using a Circle Graph

The graph shows how 300 teenagers responded to the question "How old were you when you first started using a computer?"

When Did You First Use a Computer?

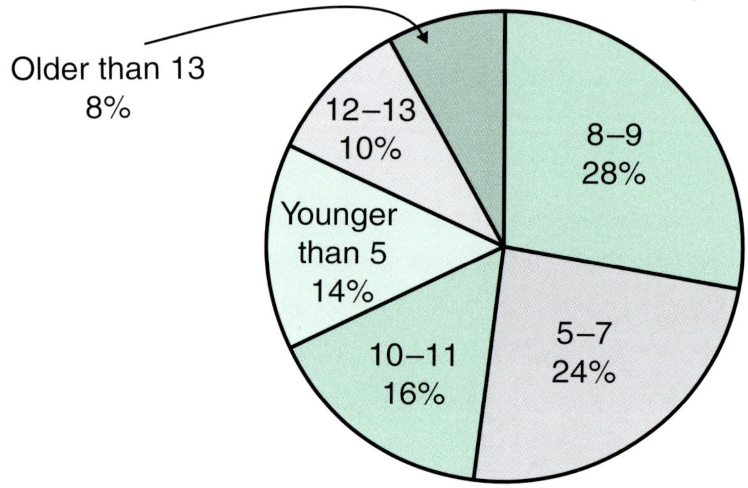

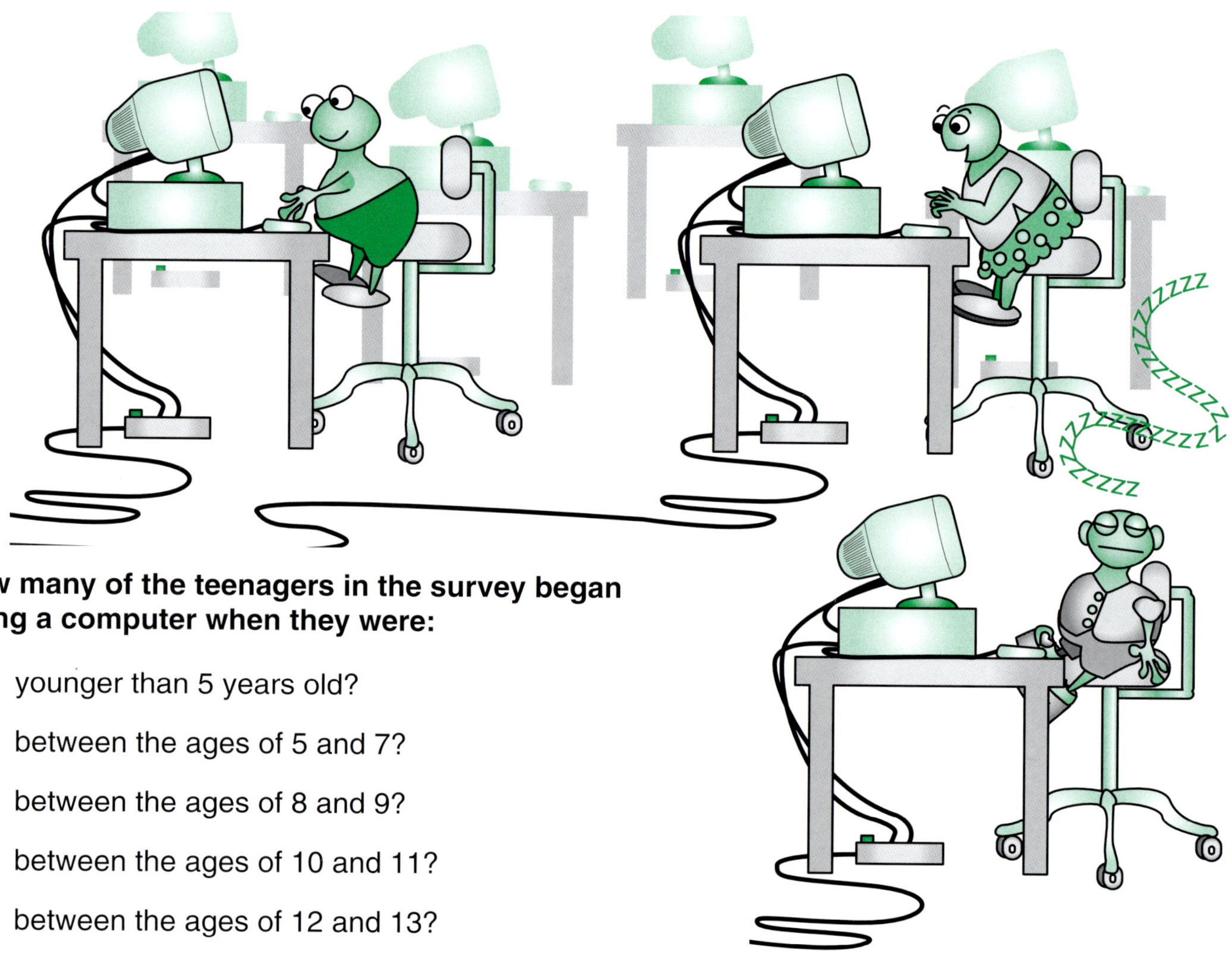

How many of the teenagers in the survey began using a computer when they were:

1. younger than 5 years old?
2. between the ages of 5 and 7?
3. between the ages of 8 and 9?
4. between the ages of 10 and 11?
5. between the ages of 12 and 13?

How many of the teenagers in the survey began using a computer when they were:

6. older than 13?
7. younger than 10?
8. between the ages of 8 and 13?
9. older than 11?

Another survey of 950 teenagers showed the same percents for the following categories: "Younger than 5" and "Older than 13." In this survey, how many teenagers said they began using a computer when they were:

10. younger than 5?
11. older than 13?
12. older than 5, but younger than 13?

Answer Box

A	B	C	D	E	F
741	72	198	42	30	54
G	**H**	**I**	**J**	**K**	**L**
24	162	48	133	84	76

Objective: Solve a problem using information from a circle graph.

And the Percent Is . . .

Example

Find the percent.

What percent of 60 is 24?

- Express "What percent" as x and "is" as the equal sign (=). "Of" means multiply.

  What percent of 60 is 24?
  $\downarrow \quad \downarrow \; \downarrow \; \downarrow \; \downarrow$

- Change the question into an equation.

  $x \times 60 = 24$

- Solve for x.

  $60x = 24$

  $60x = 24$

  $x = \frac{24}{60} = 0.4$

- Multiply the product by 100.

  $0.4 \times 100 = 40$

  So, 24 is **40%** of 60.

Find the percent.

1. What percent of 40 is 16?
2. What percent of 80 is 60?
3. What percent of 56 is 28?
4. 48 is what percent of 60?
5. 60 is what percent of 48?
6. 40 is what percent of 250?
7. What percent of 80 is 200?
8. What percent of 250 is 160?
9. What percent of 75 is 24?
10. 120 is what percent of 30?
11. Out of a shipment of 125 computers to a school, 5 did not work. What percent of the computers did not work?
12. The total cost of the computers was \$187,500. The school collected \$37,500 from the community. What percent did the community contribute?

Answer Box

A	B	C	D	E	F
50%	4%	40%	80%	16%	64%
G	**H**	**I**	**J**	**K**	**L**
125%	32%	250%	400%	20%	75%

 Objective: Use a proportion to find what percent one number is of another.